To my parents. To my family.
And to Giulio and his stories filled with music and stars.

Paola

To my ultimate creation.

Rossana

Who Will It Be?

Blue Dot Kids Press
www.BlueDotKidsPress.com

Original English-language edition published in 2020 by Blue Dot Kids Press, PO Box 2344, San Francisco, CA 94126, Blue Dot Kids Press is a trademark of Blue Dot Publications LLC.

Original English-language edition © 2020 Blue Dot Publications LLC
Original English-language edition translation © 2020 Denise Muir
Italian-language edition originally published in Italy under the title *Chi Sarâ?*
© 2018 by Camelozampa

To request permissions or to order copies of this publication, contact the publisher at www.BlueDotKidsPress.com.

Cataloging in Publication Data is available from the United States Library of Congress.
ISBN: 978-1-7331212-0-0

Printed in China with soy inks.
First Printing

Paola Vitale

Rossana Bossù

Who Will It Be?

How Evolution Connects Us All

BLUE DOT KIDS PRESS

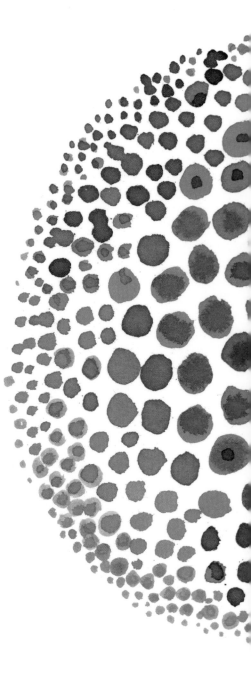

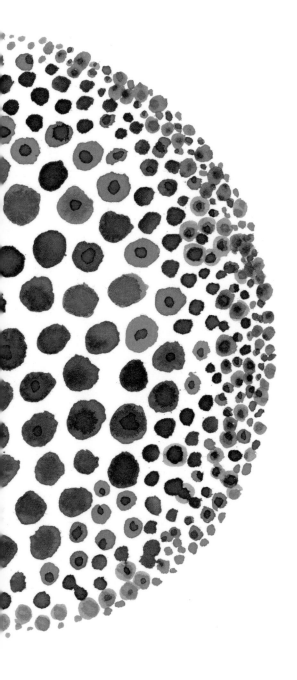

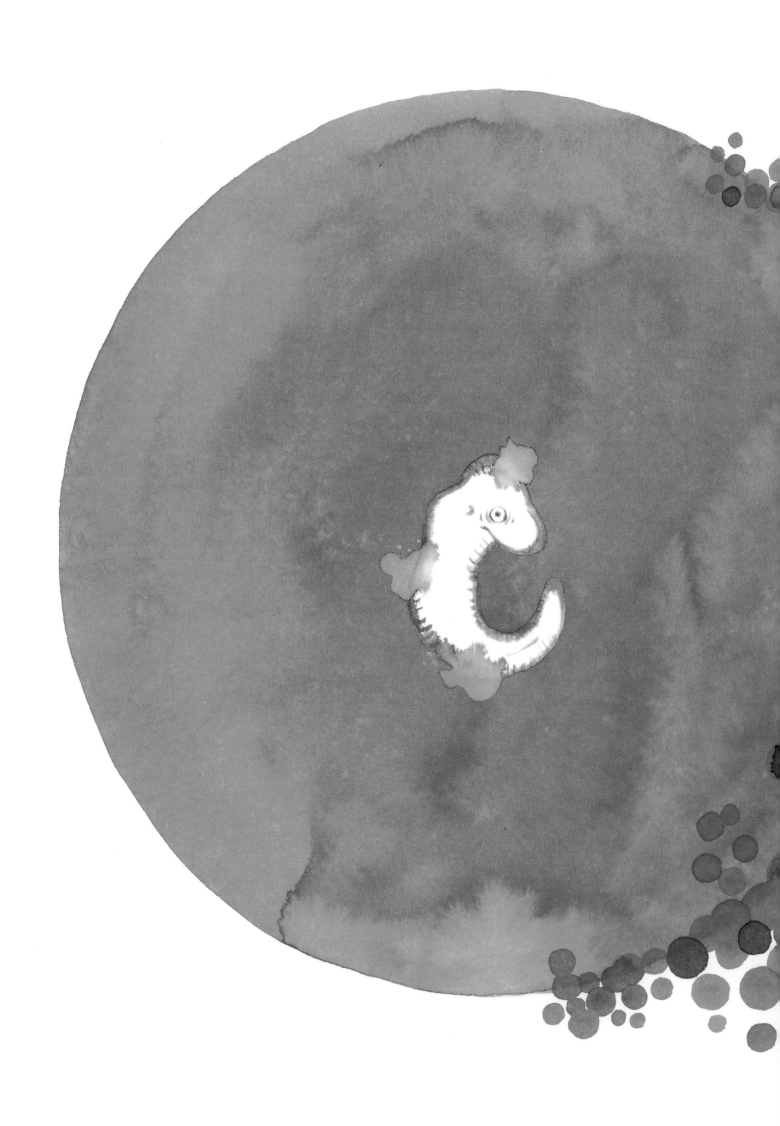

Who will it be?

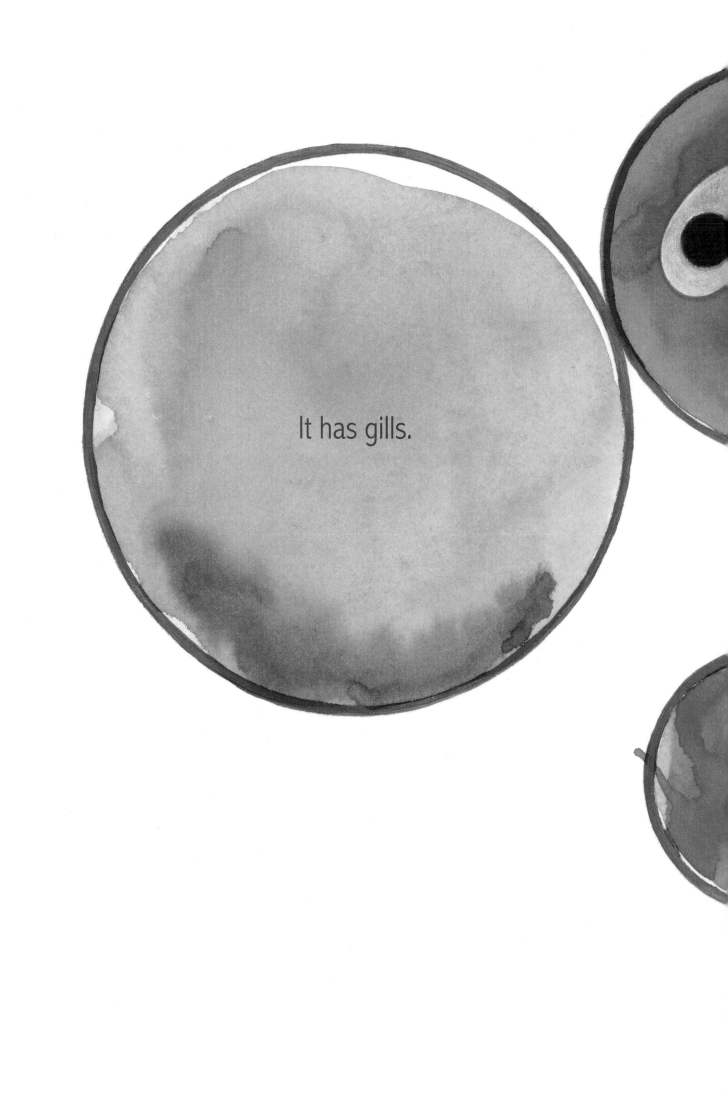

It has gills.

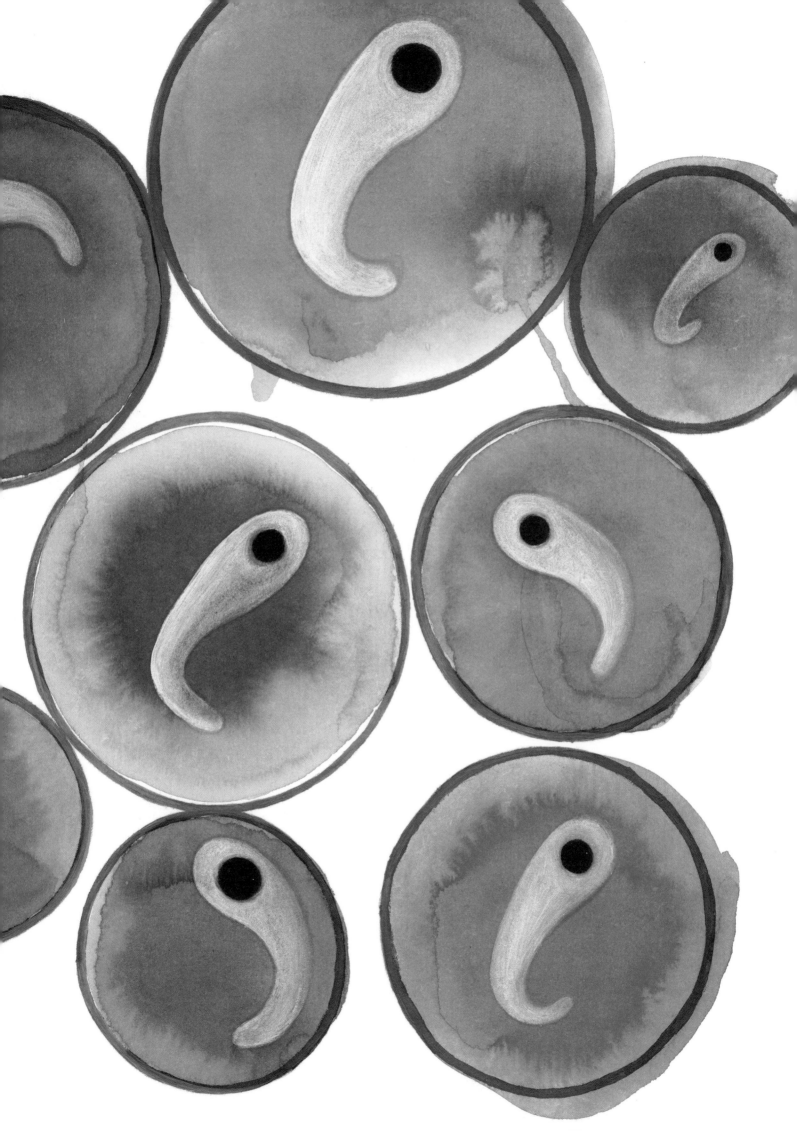

It's a fish!

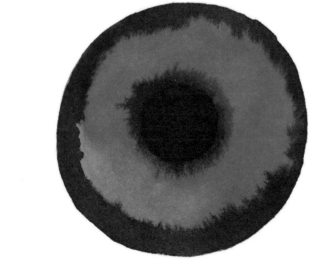

No, it loves water but has lungs.
It's an amphibian, then!

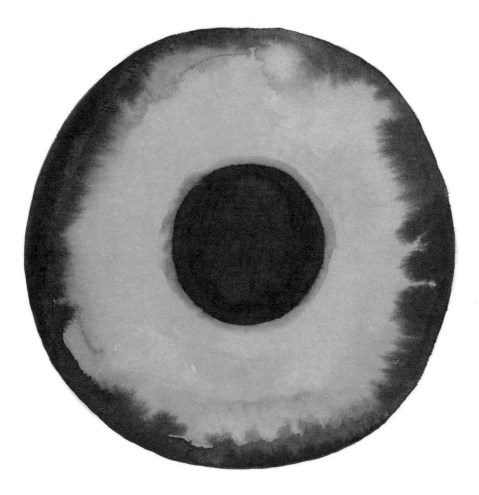

Is it a tree frog? A large toad?

It's opening its mouth wide...

Could it be a snake eating its dinner?

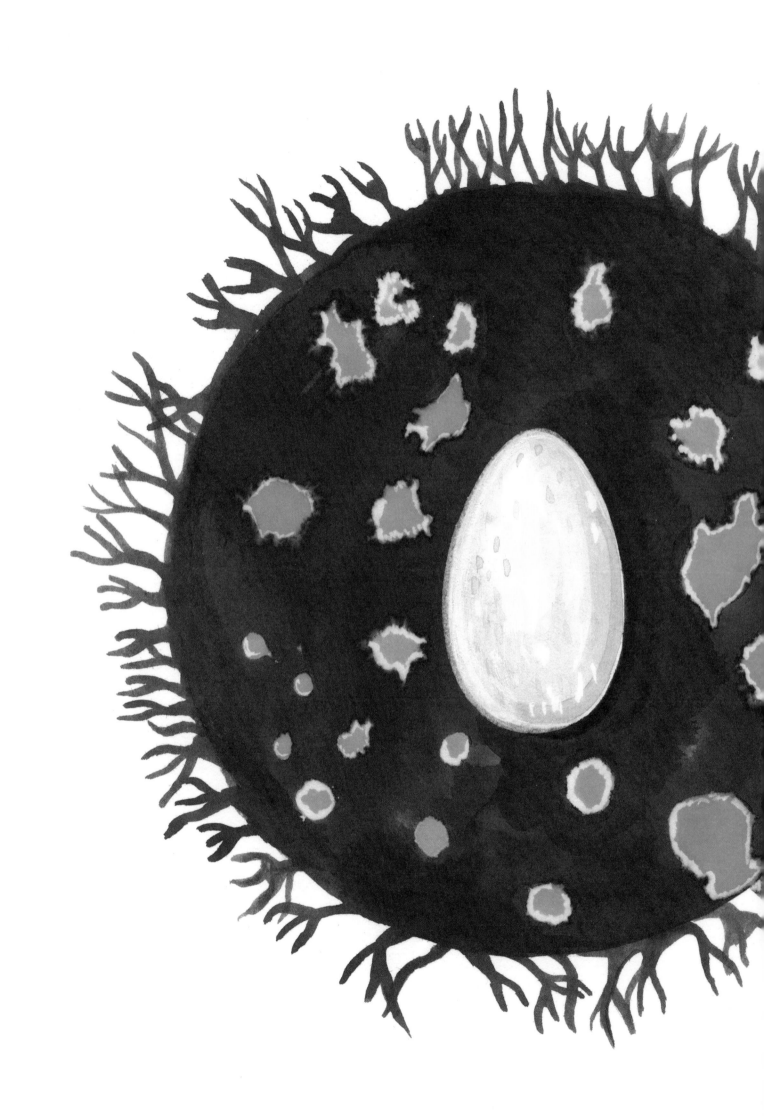

It has webbed feet.
Maybe it's a duck.

Will it fly?

It has a tail. Will it be a fox?

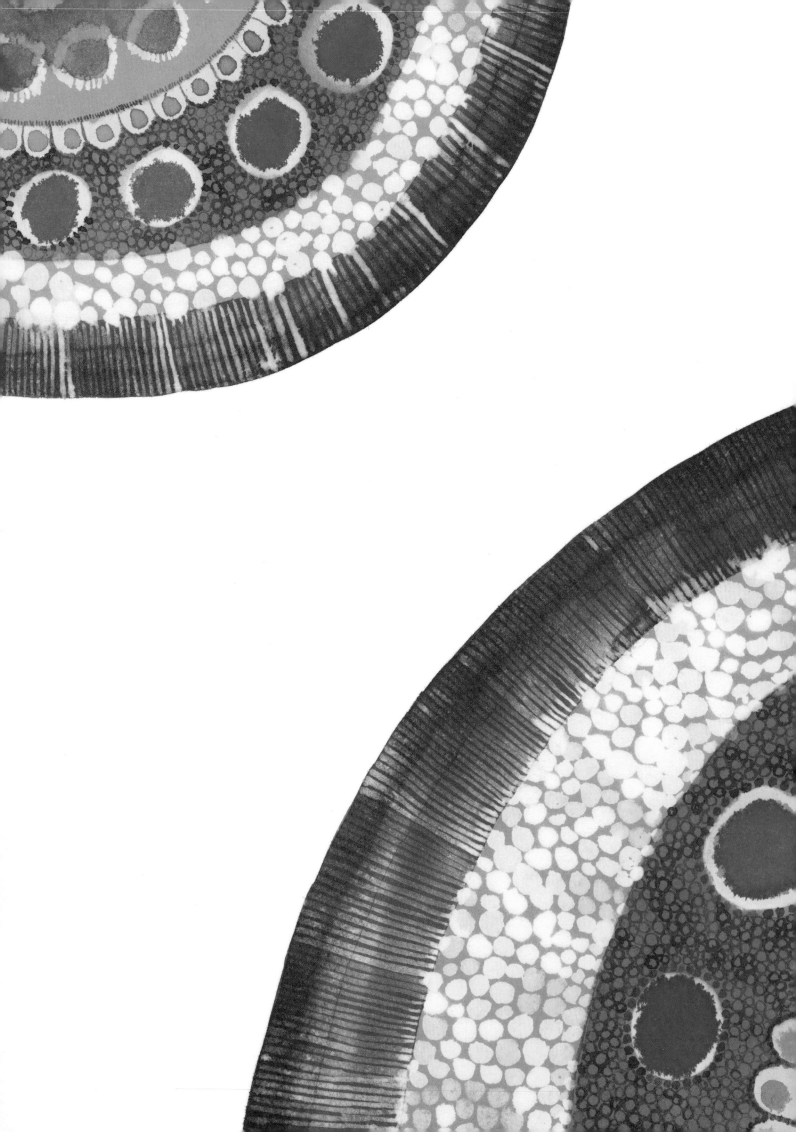

No...

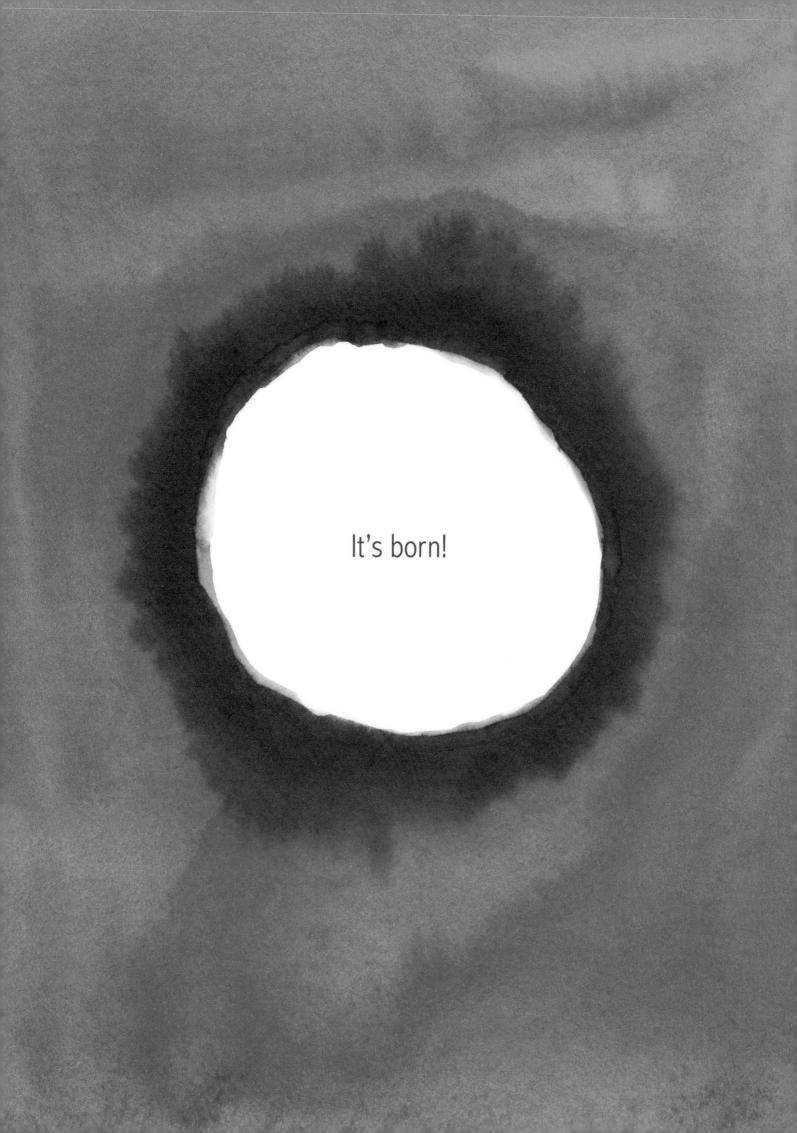

It's born!

Who is it?

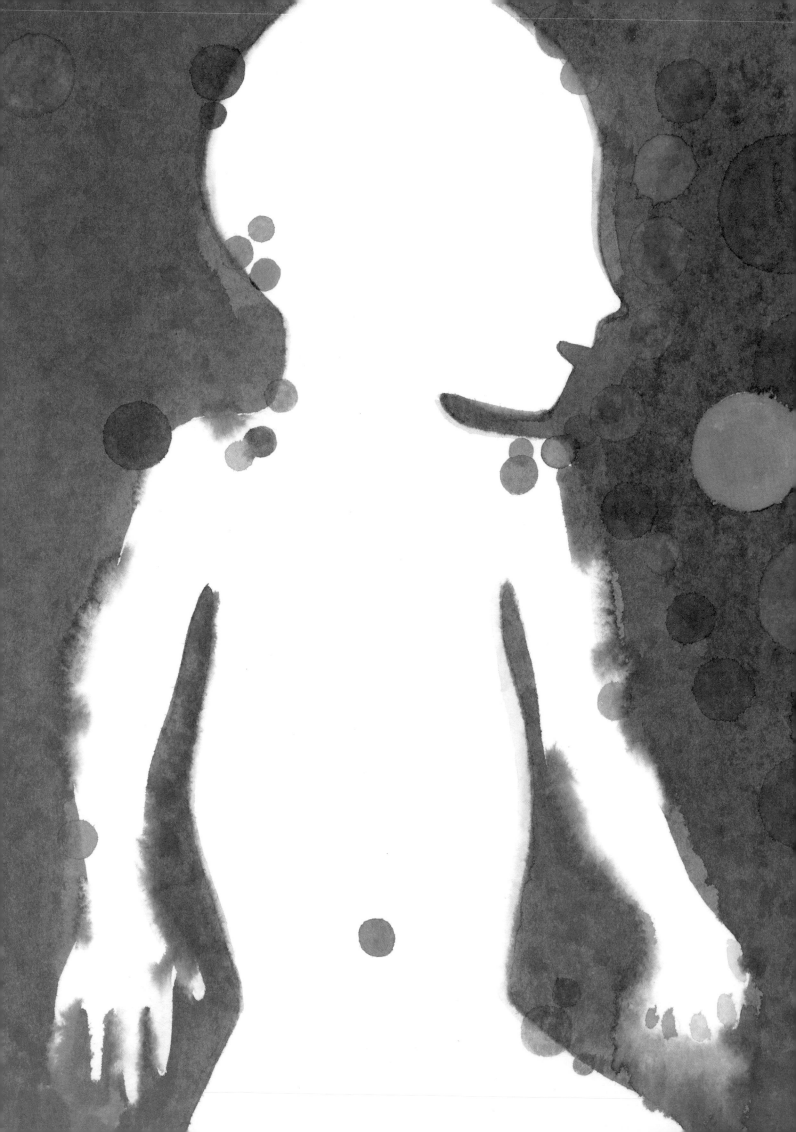

Human beings are mammals,

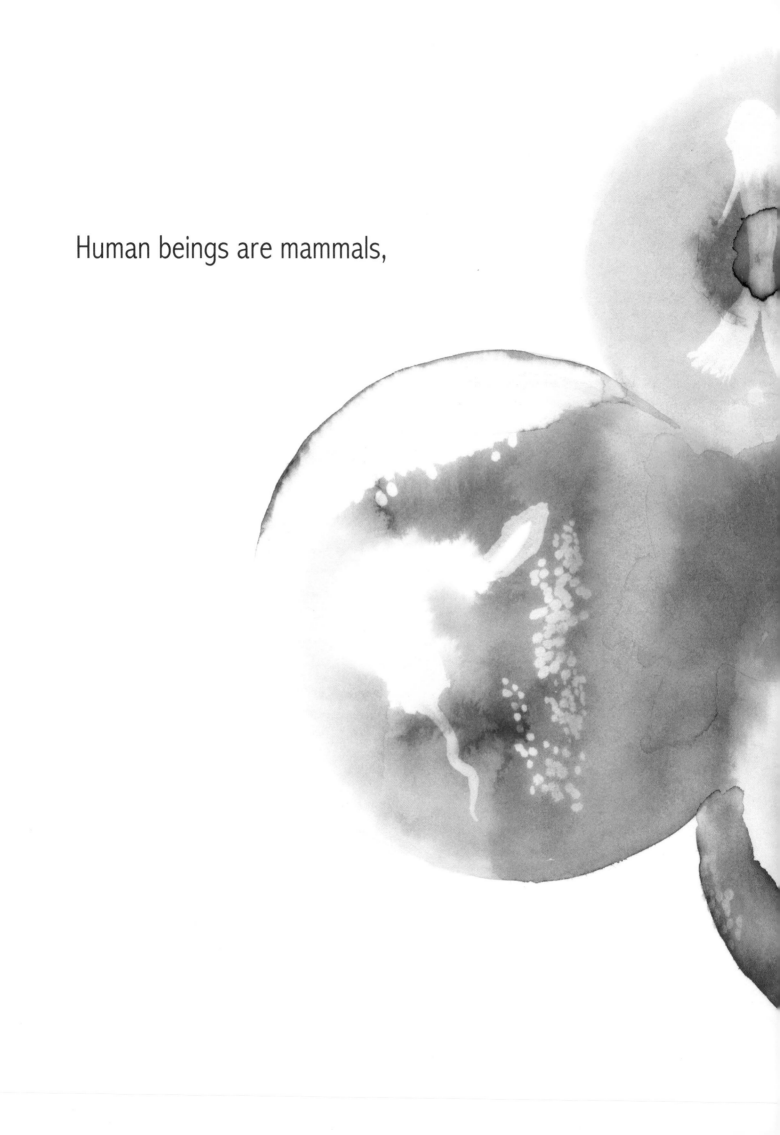

but we are also part fish,
amphibian, reptile, bird.

When we are born,
each of us already has a story to tell.

Like the memory of a long journey
stored inside each of our cells.

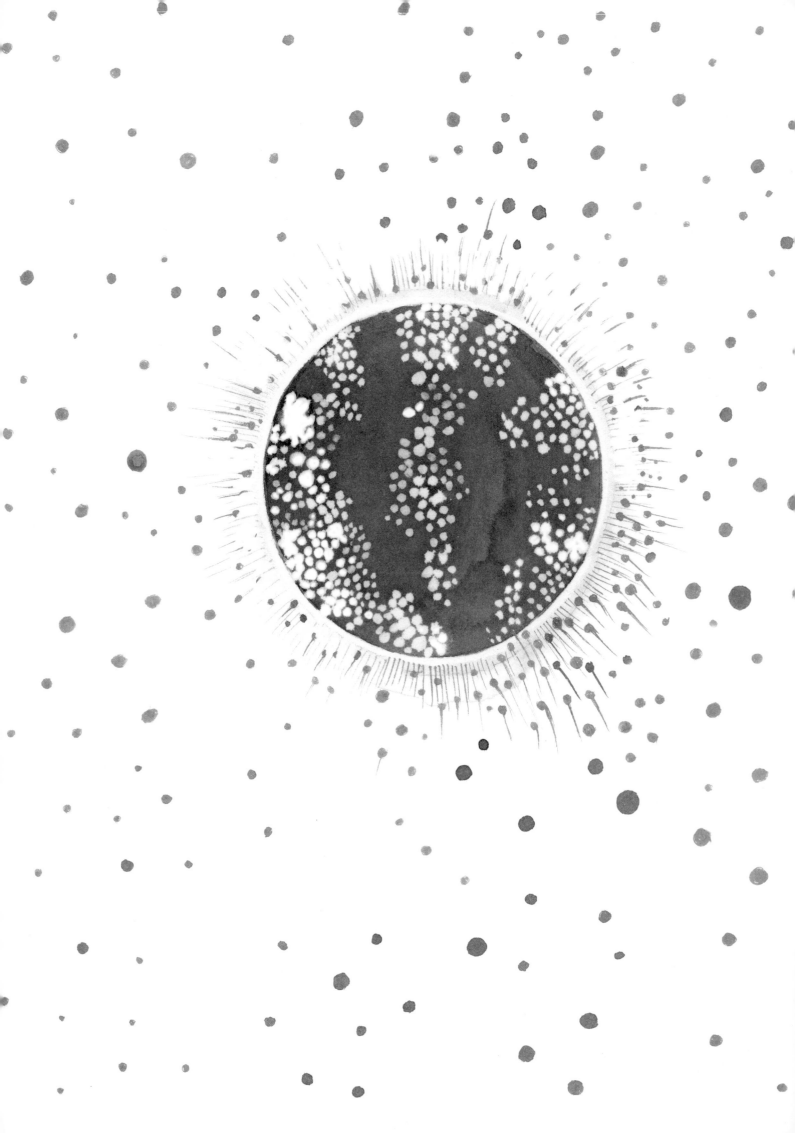

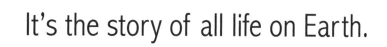

It's the story of all life on Earth.

In his 1859 book *On the Origin of Species*, Charles **Darwin** wrote that all living creatures come from simpler life forms. Everything started from single cells that lived in the oceans covering Earth four billion years ago.

Since then, life keeps changing.

Creatures that change to fit where they live survive better. This is called **natural selection**.

All of the animals we see around us today are the result of all these changes.

The earliest **vertebrates** (animals with backbones) were fish. Then came amphibians, reptiles, birds, and mammals, including humans. We all came from one common ancestor.

Humans evolved later than other vertebrates, which is why we share similarities with the animals that came before us.

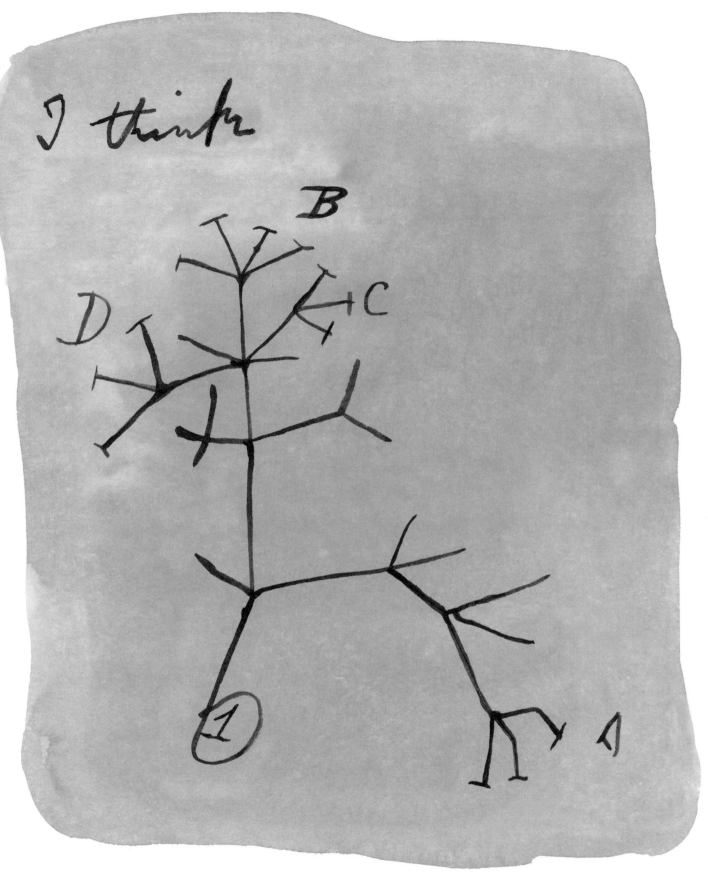

The first drawing of the Tree of Life, by Charles Darwin.

Darwin was the first person to realize that the history of life on our planet can be shown as a tree.
His **Tree of Life** has branches starting from a single trunk, which represents the first living thing on Earth.
Each branch shows the different animals that evolved later.

Proof of Darwin's idea is all around us. You can find **fossils**, the remains of once-living things, in the ground in many areas.

Fossils show us how life has changed over millions and billions of years.

In fact, we humans go on this same journey before we are born.

Life always begins with a single cell. That cell splits and grows to become an embryo. The embryo keeps growing until a new life is born. Its growth is controlled by **DNA**, a substance in each cell that carries all the important information about what the cell will become.

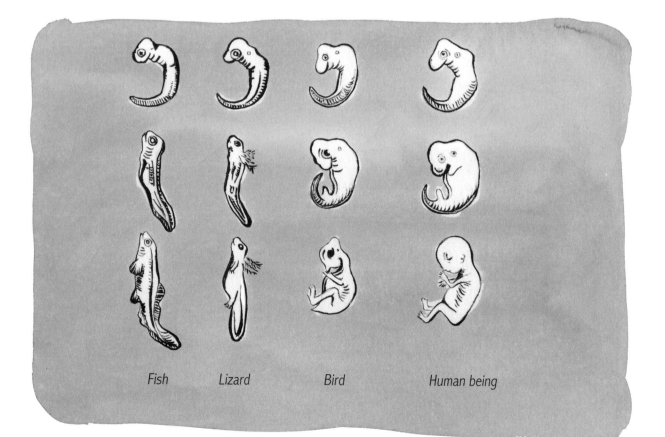

Fish   *Lizard*   *Bird*   *Human being*

In many embryos, something
very **special happens** after
development begins.

Scientists compare this event to the
narrowest part of an hourglass.
It is when differences disappear
and embryos look like they could
become the same animal.

Then the hourglass
widens again,
and embryos develop
in different ways.
But at that extraordinary
moment, it's difficult to guess
if the embryo will be a fish,
a frog, a snake, a duck
. . . or a human!

As a human embryo grows, it develops...

Slits on the neck. If it were a fish, those would become **gills** for breathing under water. Instead, the slits become lungs.

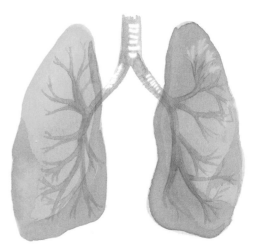

Human lungs

Webbed fingers and toes, like a duck.

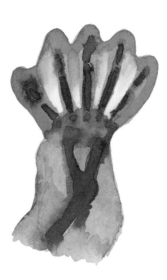

But soon those change to fingers and toes..

Two tiny bones, which lets a snake open its mouth wide. In humans, these bones become part of the ear instead.

A small tail, like a fox's. This tail becomes the end of a human spine.

These findings show Darwin's idea was right. Humans have body parts that no longer serve their original purpose.

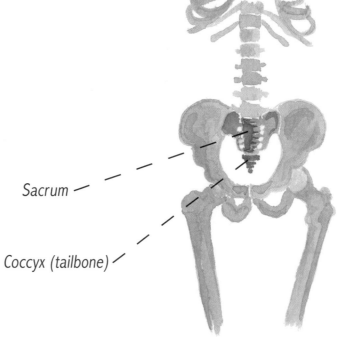

Sacrum

Coccyx (tailbone)

Looking at embryos also shows that Darwin's idea was right.

As an embryo grows, it looks like other vertebrates.
Our embryos look similar because we all came from a single ancestor.
In fact, we look more similar to animals we are more closely related to, or closer to on the Tree of Life.

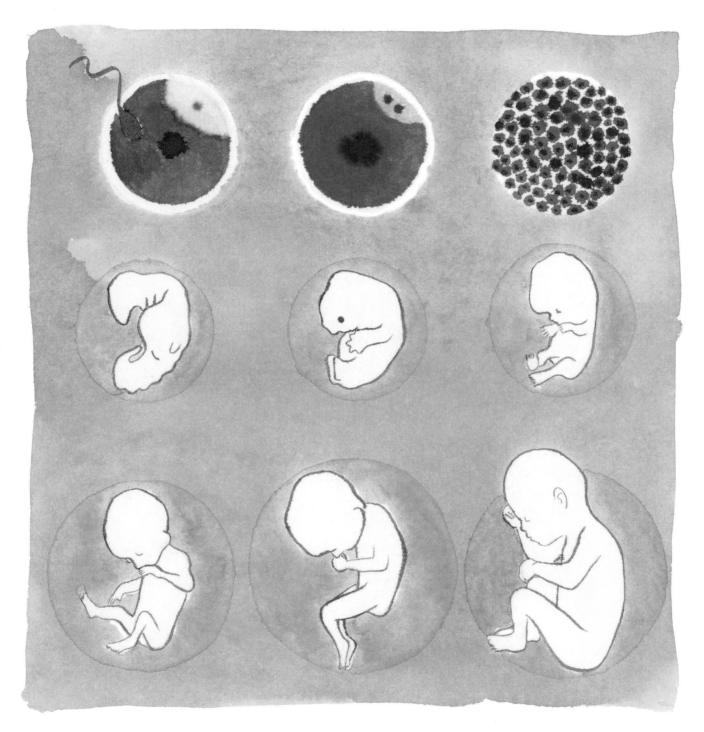

**Before being born, it's like every baby "remembers"
the journey that life took before humans appeared.**

More than a century after *On the Origin of Species*, scientists studying DNA confirmed Darwin's amazing, revolutionary idea.

There is an extraordinary moment when all embryos are similar because we all share ancient **DNA** from the first vertebrates on Earth.

We are most similar to chimpanzees, but we have things in common with zebrafish and even the nematode worm!

We humans have something in common with all living creatures.

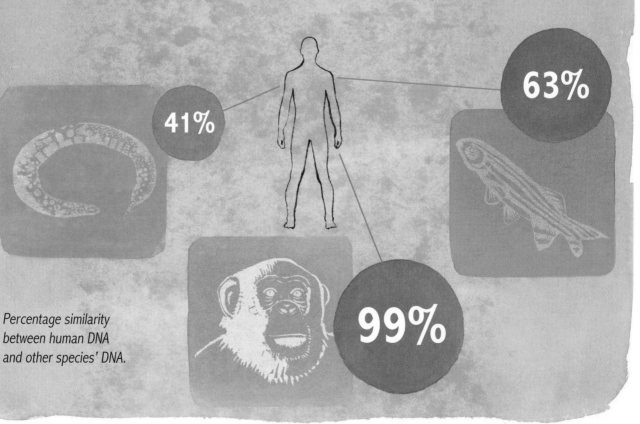

41%

63%

99%

Percentage similarity between human DNA and other species' DNA.

AUTHOR

Paola Vitale graduated with a degree in biological sciences from the University of Padua, Italy. She has previously published international scientific studies and has a PhD in developmental biology. She is a teacher and a children's author.

ILLUSTRATOR

Rossana Bossù lives in Turin, Italy. She is a former teacher at the Art and Design Institute in Turin and has won several prizes as an illustrator. She is the author and illustrator of such best sellers as *How Big Is an Elephant?*, which has been translated into ten languages.

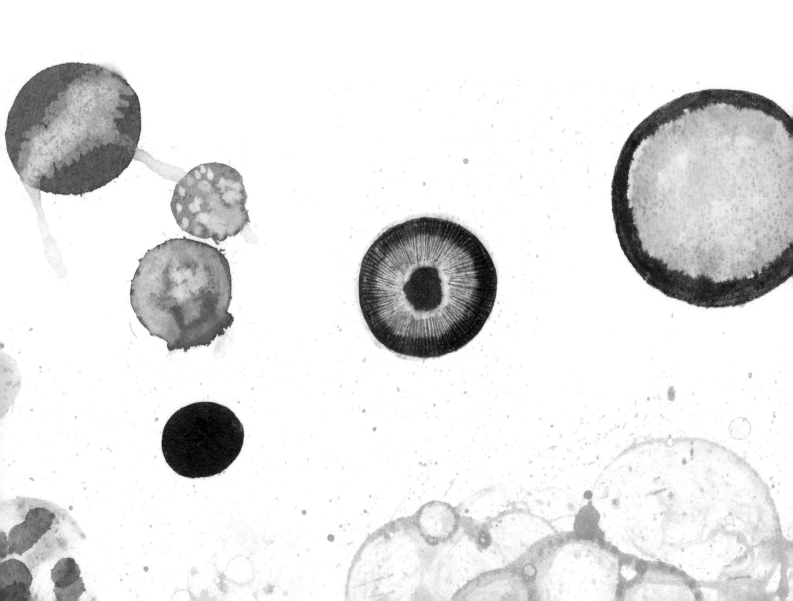

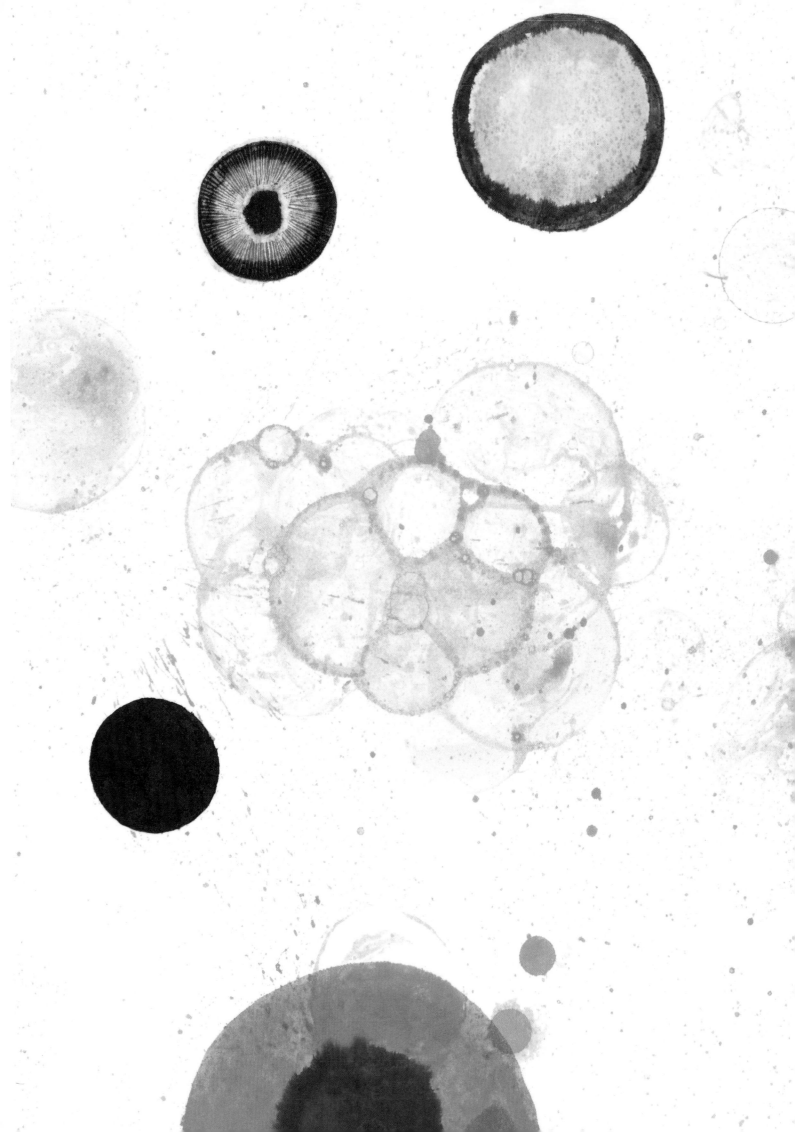